Nuclear Engineering Division

AA009176

Held at the 2024 AIChE Annual Meeting

San Diego, California, USA
27-31 October 2024

ISBN: 979-8-3313-1679-2

Printed from e-media with permission by:

Curran Associates, Inc.
57 Morehouse Lane
Red Hook, NY 12571

Some format issues inherent in the e-media version may also appear in this print version.

Copyright© (2024) by AIChE
All rights reserved.

Printed with permission by Curran Associates, Inc. (2025)

For permission requests, please contact AIChE
at the address below.

AIChE
120 Wall Street, FL 23
New York, NY 10005-4020

Phone: (800) 242-4363
Fax: (203) 775-5177

www.aiche.org

Additional copies of this publication are available from:

Curran Associates, Inc.
57 Morehouse Lane
Red Hook, NY 12571 USA
Phone: 845-758-0400
Fax: 845-758-2633
Email: curran@proceedings.com
Web: www.proceedings.com

TABLE OF CONTENTS

GRADUATE STUDENT AND EARLY CAREER INVESTIGATIONS - SESSION I

SEPARATION ADVANCES IN NUCLEAR CHEMICAL PROCESSING

THEORY, MODELING, AND SIMULATION OF NUCLEAR CHEMICAL PROCESSES

NONPROLIFERATION STEWARDSHIP SUPPORT FOR U/PU PROCESSING

RADIOCHEMICAL PROCESSING

NUCLEAR INDUSTRY AND NATIONAL CHALLENGES

NUCLEAR CHEMICAL ENGINEERING OUTLOOK (INVITED TALKS)

GRADUATE STUDENT AND EARLY CAREER INVESTIGATIONS - SESSION II

146a Mixed Extractant Systems for Extraction Chromatography

Adam Boey, Erin Bertelsen

The lanthanides and actinides play crucial roles in nuclear medicine, nuclear energy production, and clean energy technologies, necessitating advanced separation processes for their extraction and purification. Modified extractant systems have shown enhanced metal recovery[1-3] and/or reduced undesirable phase phenomena using solvent extraction (SX) techniques.[4-6] Extraction chromatography (EXC), a subset of liquid-liquid chromatography, employs a stationary phase of a neat (undiluted) metal extractant, or a metal extractant dissolved in a diluent, physisorbed on porous solid support. The extractants commonly used in EXC are typically the same as those used in SX systems, granting EXC the selectivity of SX alongside the multiple separation stages and operational simplicity of chromatography. However, synergism has not been extensively explored using this separation technique.

In this study, two *f*-element extractants, 2-ethylhexyl 2-ethylhexylphosphonic acid (HEH[EHP]) and tetra-(2-ethylhexyl)diglycolamide (TEHDGA), were physisorbed on a polymeric porous solid support and evaluated for Eu(III) uptake using batch extraction techniques. The combined extractant system exhibited synergism from 0.14–0.96 M HNO_3 but was antagonistic at lower acid concentrations. Furthermore, the TEHDGA + HEH[EHP] EXC material provided a higher maximum Eu(III) adsorption capacity (155 ± 3 μmol g^{-1}) compared to the single extractant HEH[EHP] system (85 ± 8 μmol g^{-1}).

1. Dyrssen, D.; Kuča, L. The Extraction of Uranium(VI) with DBP in the Presence of TBP. The "Synergic" Effect: Substitution or Addition? *Acta Chem Scand.* **1960**, *14*, 1945–1956, DOI: 10.3891/acta.chem.scand.14-1945.
2. Hall, G. B.; Holfeltz, V. E.; Campbell, E. L.; Boglaienko, D.; Lumetta, G. J.; Levitskaia, T. B. Evolution of Acid-Dependent Am^{3+} and Eu^{3+} Organic Coordination Environment: Effects of the Extraction Efficiency. *Chem.* **2020**, *59*, 4453–4467, DOI: 10.1021/acs.inorgchem.9b03612.
3. Dashti, S.; Sadri, F.; Shakibania, S.; Rashchi, F.; Ghahreman, A. Separation and solvent extraction of rare earth elements (Pr, Nd, Sm, Eu, Tb, and Er) using TBP and Cyanex 572 from a chloride medium. *Eng.* **2021**, *161*, 106694, DOI: 10.1016/j.mineng.2020.106694.
4. Vasudeva Rao, P. R.; Kolarik, Z. A Review of Third Phase Formation in Extraction of Actinides by Neutral Organophosphorus Extractants. *Solvent Extr. And Ion Exch.* **1996**, *14*, 955–993, DOI: 10.1080/07366299608918378.
5. Gujar, R. B.; Ansari, S. A.; Prabhu, D. R.; Pathak, P. N.; Sengupta, A.; Thulasida, S. K.; Mahapatra, P. K.; Manchanda, V. K. Actinide Partitioning with a Modified TODGA Solvent: Counter-Current Extraction Studies with Simulated High Level Waste. *Solvent Extr. And Ion Exch.* **2012**, *30*, 156–170, DOI: 10.1080/07366299.2011.609392.
6. Woodhead, D.; MaLachlan, F.; Taylor, R.; Müllich, U.; Geist, A.; Wilden, A.; Modolo, G. Nitric Acid Extraction into a TODGA Solvent Modified with 1-Octanol. *Solvent Extr. And Ion Exch.* **2019**, *37*, 173–190, DOI: 10.1080/07366299.2019.1625201.

146b Mixer Pump Operations for Hanford Waste Tanks

Megan Higley, Sarah Suffield, Beric Wells, Courtney Bottenus

The Hanford Site's double-shell tanks (DSTs) contain supernatant and undissolved solid waste from previous plutonium production. This waste needs to be removed from the tanks and immobilized through vitrification. For efficient vitrification, a relatively homogenous feed with specified solid concentrations from the tanks is required. An in-tank settling and decanting process can be used to concentrate the solid waste. After being concentrated, the remaining liquid would be used to mobilize and suspend the solid particles with mixer pumps for removal from the tanks with a transfer pump. Computational fluid dynamics (CFD) simulations can be used to investigate the settling/decanting process and optimize the mixer and transfer pump specifications and conditions of operation. The CFD simulations provide key input to the planned scaled experimental testing and scale-up methods that will be used to demonstrate, verify, and provide the final recommendations for the equipment configuration and operation of the system. A CFD model of the waste tank has been constructed which includes the rotating mixer pumps with variable nozzle sizing. In this work, the CFD model is used to optimize mixing performance by evaluating the effects of mixer pump nozzle diameter and flowrate on upward flow velocity in specific regions of the tank.

146c Evaluation of High Level Waste Sludge Processing Behavior

Austin Bachman, Amy Westesen, Kyleigh Murray, Carolyn A. Burns, Reid Peterson

The U.S. Department of Energy's (DOE) Hanford Site has 177 underground storage tanks that contain a complex and diverse mix of chemical and radioactive wastes from past nuclear fuel reprocessing and waste-management operations. The strategy of the DOE Office of River Protection is to retrieve and treat this waste. Approximately 20 vol% of this waste is in the form of insoluble undissolved solids (UDS) or sludge, and this waste may require modification prior to its transfer for treatment at the Waste Treatment and Immobilization Plant (WTP). Required modification may be due to specific WTP acceptance criteria for waste feed delivery which describe the physical and chemical properties of the waste that must be met before the waste is transferred to the WTP. Modification may also occur from incremental and incidental waste blending during retrieval from the individual waste tanks. One challenging requirement relates to the UDS composition in a waste feed because the solid particles settle, and their concentration and relative proportion can change during batch transfers of the waste to WTP. A key uncertainty is the ability to mix and transfer wastes with large variations in UDS concentrations and resulting settling rates. To address this uncertainty, a number of bench-scale mixing and settling tests have been conducted to determine the process performance for variable physical and chemical properties of tank waste sludge and simulant samples. Settling velocity, chemical characterization, shear strength measurements, and just suspended impeller speed (N_{js}) were determined for tank waste simulants and actual waste sludges to evaluate the mobilization of non-cohesive particles at various laboratory scales. This presentation details a comparison of the size and density of sludge particulate between simulants and actual waste sludge to help understand the behavior of high-level waste solids settling and transport. Variations in sludge chemistry are used to determine the possibility of potential solids segregation and uneven particle distribution. Understanding the behavior of high level waste sludge transport can provide potential remedies for addressing problematic behavior and technical guidance during high-level waste flowsheet planning.

146d Combining Spectroscopic Measurements with Mass-Balance Predictions Via the Dual Kalman Filter for Real-Time Monitoring

Steven Crouse, Ronald Rousseau, Martha Grover

The Hanford Site in Washington State contains approximately 56 million gallons of nuclear and chemical waste stored in 177 underground tanks [1]. To prevent future environmental contamination, a permanent storage solution is sought. Current plans from the United States Department of Energy call for the waste from these tanks to be separated into low activity waste (LAW) and high-level waste (HLW) streams before being stabilized through vitrification. There are three leading methods for estimating the compositions of waste streams as they undergo processing and, ultimately, vitrification at the Hanford Site: grab sampling, real-time measurements, and mass-balance predictions. However, there has been little reported research combining all three methods for accurate and efficient composition estimation and fault detection. The present research focuses on the practical aspects of "what can go wrong in waste processing at the Hanford site?" and how such situations can be identified in a timely and accurate manner. This work particularly focuses on a variable waste feed stream and potential changes in mixing within the tanks.

In the context of the Hanford Site, the three methods used to estimate compositions have distinct advantages and disadvantages.

- Grab-sampling allows accurate and complete characterization of waste samples, but also has risks associated with potential exposure of workers to radiation and it can be a rate-limiting-step in nuclear-waste processing.
- A mass-balance model instantaneously produces estimates of compositions that only require up-stream information. However, a mass-balance approach provides a static model and has limited adaptability to variable process conditions.
- Real-time monitoring provides instantaneous process feedback making measurements robust to changing process conditions. However, not every species can be detected with every spectroscopic technique, limiting the number of species measured directly. Additionally, there are practical challenges such as window probe fouling.

The goal of the present work is to combine grab-sampling, a mass-balance model, and real-time measurements to create a stream characterization framework that is robust, efficient, and accurate.

The methodology is developed using a simulation of a representative Hanford-like system. The simulation is comprised of a 3-tank batch system modeling the Concentrate Receipt Vessel, Melter Feed Preparation Vessel, and Melter Feed Vessel (shown in Figure 1). In addition, attention has been given to modeling aspects unique to the Hanford site such as mixing within the melter and sample acquisition time. For the simulation, time steps forward from an initial condition with uncertainty being introduced with every tank transfer and with the addition of glass-forming chemicals. At each time-step, the contents of one tank are emptied into the next tank, except for residuals (also known as heels) that are left behind [2]. Assumptions are made on which species can be readily measured with ATR-FTIR and Raman spectroscopies (and their respective measurement accuracies) based on prior experimental work.

A state- and parameter-estimation technique, the dual-Kalman Filter, is used in this study to combine grab-sampling and real-time measurements with a mass-balance model for chemical composition estimates and real-time parameter estimates of the mass-balance model [3]. Operationally, the filter can detect faults by comparing the measured mass-balance model to a nominal (expected) mass-balance model. When the heel-mass of the measured mass-balance model deviates from the heel-mass expected from the nominal mass balance model, a fault is suspected. In the context of Hanford and the current study of heel mass mixing, a fault may indicate that the stirring mechanism is broken, crystallization/precipitation of a species has occurred within a tank, there is buildup on the tank walls, or the size or morphology of insoluble solids in the waste stream have changed. Regardless of the cause of the fault, engineers should be alerted of the changed process conditions and assess the cause before a bad batch of glass is produced or equipment (such as the melter) sustains damage.

The dual-Kalman Filter, in general, is used to combine real-time measurements with a mass-balance model predictions and grab-sampling. The dual-Kalman filter and its varieties (Kalman filters) are well-established and have been applied to many monitoring tasks over the past 50 years. The dual-Kalman filter combines data sources by weighting them based on their uncertainty. Because of the incorporation of uncertainty into the model, error bounds can be reported in addition to the predictions themselves.

Vitrification of nuclear waste at the Hanford site is a large undertaking that will take decades and a hundreds of billions of dollars. Because of the substantial investment of capital and time, effective and fault-free waste processing is highly desirable to avoid prolonging remediation efforts. This study showcases how the dual-Kalman filter can integrate a mass balance model with grab-sample measurements and in-line measurements to detect faults in heel mass composition. Furthermore, the dual-Kalman filter in this work highlights how real-time measurements may be utilized with existing process models at Hanford to incorporate real-time information for process monitoring. Process faults are not planned, are difficult to predict, and cannot always be prevented. Therefore, fault detection at the Hanford site will have to rely on observation and measurement.

Figure 1. Different measurement scenarios of the three-tank system studied in this work (attached image).

References

[1] R. A. Peterson *et al.*, "Review of the Scientific Understanding of Radioactive Waste at the U.S. DOE Hanford Site.pdf," *Environ. Sci. Technol.*, no. 52, pp. 381–396, 2018.

[2] G. F. Piepel, B. G. Amidan, A. Heredia-langner, D. R. Weier, and S. K. Cooley, "Statistical Methods and Results for WTP IHLW and ILAW Compliance," 2005.

[3] T. A. Wenzel, K. J. Burnham, M. V. Blundell, and R. A. Williams, "Dual extended Kalman filter for vehicle state and parameter estimation," *Veh. Syst. Dyn.*, vol. 44, no. 2, pp. 153–171, 2006, doi: 10.1080/00423110500385949.

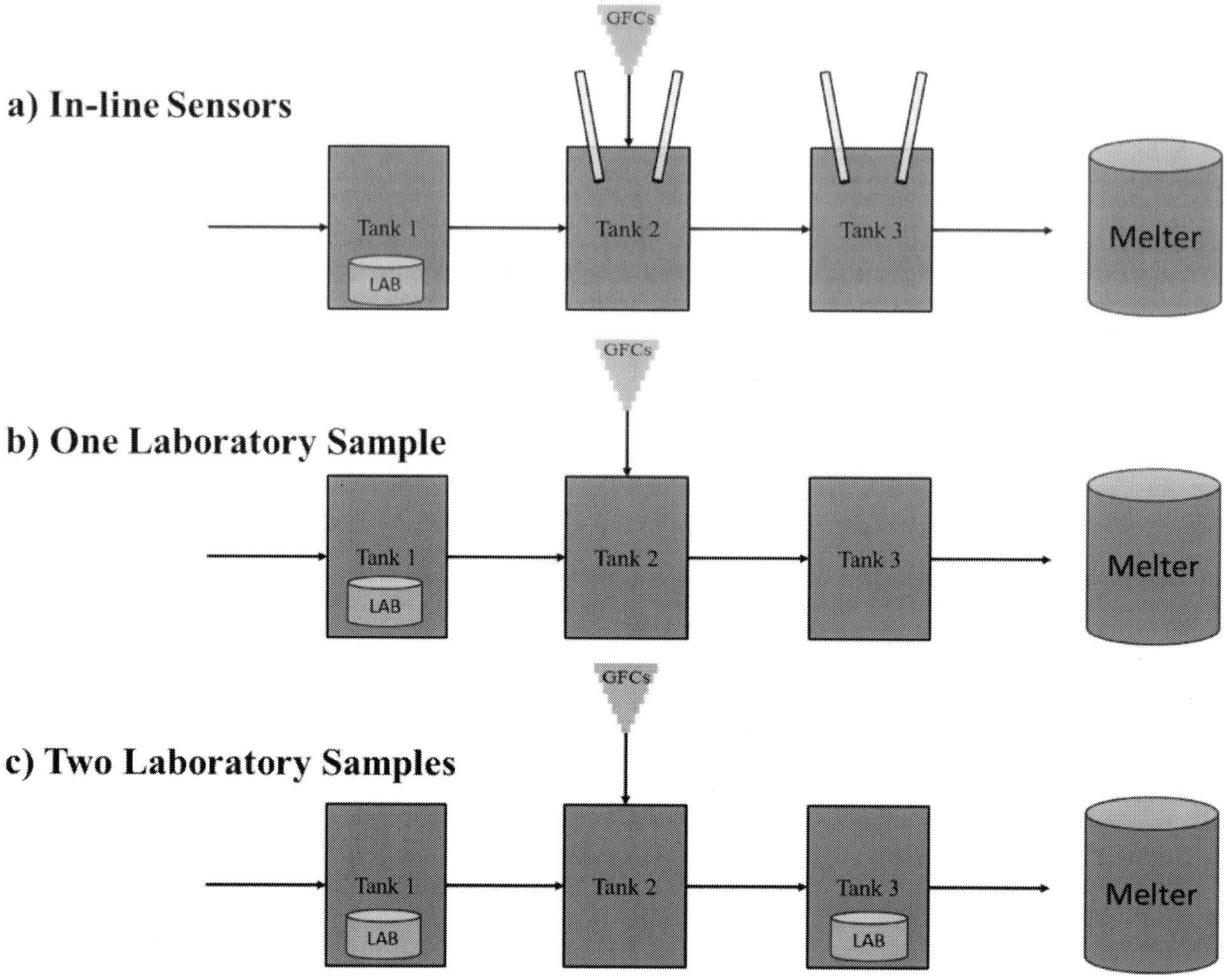
a) In-line Sensors
GFCs
Tank 1
LAB
Tank 2
Tank 3
Melter
b) One Laboratory Sample
GFCs
Tank 1
LAB
Tank 2
Tank 3
Melter
c) Two Laboratory Samples
GFCs
Tank 1
LAB
Tank 2
Tank 3
LAB
Melter

146e Chemistry Control and Hydrogen Speciation in Molten Fluoride Salts

Haley Williams, Raluca O. Scarlat

Molten fluoride salt may be used as a fuel solvent or coolant in advanced fission reactors or may serve as a tritium breeding blanket in fusion reactors. Characterizing the salt's chemistry involves measurement of its redox potential (the oxidation state in which its species exist dictates which elements are thermodynamically favored as fluorides or as reduced species) and its fluoroacidity (the degree of association/dissociation of the ions describes how solutes are incorporated into the salt). Evaluating how the salt chemistry evolves over the course of a reactor's lifetime is of engineering concern in order to avoid corrosion, detect leaks, and ensure thermophysical properties are appropriately maintained for sufficient cooling and safe operation.

In both fission and fusion reactors which use lithium fluoride, tritium (hydrogen-3) may be generated via neutron activation of lithium-6. (It may also exist as a fission product in fission reactors.) Management of tritium inventories is a key engineering concern due to its radioactivity and ability to diffuse through metals. The species in which tritium exists in a molten fluoride system is dependent on its redox potential. Hypothetically, speciation would also be dependent on the fluoroacidity of the system, but the relationship between oxidation state, species, and solubility of tritium in molten fluorides is still unknown. In this presentation, experimental investigations of hydrogen, serving as a surrogate for tritium, in molten $2LiF\text{-}BeF_2$ will be shared within the context of chemistry control and tritium management in advanced nuclear reactors. Hydrogen species have been introduced to $2LiF\text{-}BeF_2$ via additions of LiH and via electrochemical desorption of H_2 gas from graphite. Cyclic and square wave voltammograms have been collected and will be analyzed to hypothesize the electroactive hydrogen species dissolved in the melt. Possible hydrogen species and their relationship to the redox and acidity conditions in the melt will be discussed.

Overall, these fundamental studies may inform electrochemical sensor design for chemistry control systems. Additionally, a more complete understanding of speciation is relevant to techniques for the separations or removal of tritium. A deeper understanding of the relationship and utility of speciation, structure, and electrochemistry in molten salts is also relevant to the development of pyroprocessing techniques and other technologies which employ molten salts. This talk will include a presentation of some cross-cutting studies of molten salt speciation across various technologies.

608a Kinetic Comparisons of Cesium Ion Exchange Performance with Crystalline Silicotitanate

Reid Peterson, Amy Westesen, Andrew Carney, Sandra Fiskum, Truc Trang-Le, Matt Landon

Crystalline silicotitanate (CST) ion exchanger is currently selected to remove cesium (^{137}Cs) from the aqueous phase of Hanford tank wastes. The Advanced Modular Pretreatment System (AMPS) will provide pretreatment capabilities for tank waste supernate originating from double-shell tanks in the 200 East area of the Hanford site. To optimize and evaluate exchange kinetics during AMPS cesium removal, Pacific Northwest National Laboratory conducted laboratory-scale ion exchange column tests at various residence times and scales to determine throughput rates and CST usage. Testing conditions were scaled up from laboratory to full-scale operations to assess AMPS processing conditions at the various residence times. Based on these results, a determination of mass transfer coefficients as well as tradeoffs of CST usage with flowrate were conducted. Data presented will allow for future modeling of breakthrough performance at a range of processing conditions necessary for AMPS flowsheet development.

608d Investigating Fission Product Separation from High-Level Waste Chloride Salt Systems, with a Focus on Cesium, through Melt-Crystallization Techniques.

Maria Del Rocio Rodriguez Laguna, Kevin R. Tolman, Morgan T. Kropp, Jacob A. Yingling, Stephanie Castro Baldivieso, Steven Herrmann, Tae-Sic Yoo

Nuclear fission reactors employ fissile materials, like U-235, to initiate controlled chain reactions, resulting in the generation of heat. Reprocessing of spent nuclear fuel plays a critical role in recovering fissile materials and reducing the volume of nuclear waste. One such reprocessing technique, pyroprocessing, involves dissolving spent nuclear fuel in a bath of molten salt and electrochemically separating the actinides using an electrorefiner. Typically, a LiCl-KCl eutectic salt system serves as the electrolyte and solvent. The fission products remain in the salt. Historically, the salt, along with the fission products, has been disposed of as solid waste. However, separating fission products from the salt is advantageous environmentally and economically, as it reduces the volume of waste produced. The present research proposes a novel, thermally controlled solid-liquid separation process for recovering Cs^+ from a chloride-salt matrix (LiCl-KCl-CsCl-NaCl). Cesium-137 is a high-yield fission product, with a half-life of 30 years, and is a significant source of radioactive environmental contamination with high water solubility and mobility. The choice of the salt matrix, particularly the presence of NaCl can be explained by the fact that Idaho National Laboratory has a history of processing sodium-bonded fuels. The system was investigated using differential scanning calorimetry (DSC), high-temperature X-ray diffraction (HT-XRD), inductively coupled plasma optical-emission spectroscopy (ICP-OES), and thermodynamic modeling. The calculation and analytical results showed that Cs^+ can be concentrated in the liquid phase during melt-crystallization. A novel two-heat-source melt-crystallizer was designed and used to collect liquid samples during the partial-crystallization process to demonstrate the concept.

A different method, employing a cold finger, has also been tested using this composition. Findings from the experiments corroborate the trends observed with the two-heat-source setup, with cesium being concentrated in the liquid phase. Notably, the cold finger setup offers an enhancement: it incorporates liquid mixing, facilitating the uniform distribution of temperature throughout the bulk molten salt, thus aiding in achieving thermal equilibrium more effectively.

608c Thermophysical Property Measurement of Uranium Containing Fluoride and Chloride Salts

D. Nathanael Gardner, Zachary Falkowski, Raluca Scarlat

Actinide-containing fluoride and chloride molten salts are utilized as fuel salts in molten salt reactors (MSR's) and have applications in pyrochemical-processing for fuel recovery. However, the thermophysical property data of these salts such as melting point, viscosity, and density and the effects on these properties of varying levels of actinide content are relatively unknown. This presentation will present novel data on the viscosity, density, and melting point of two candidate salts for use in MSR's, $NaCl$-UCl_3 and NaF-BeF_2-UF_4. Additionally, to model the effects of plutonium produced by fission on the fuel salts' thermophysical properties, the effects of cerium added as $CeCl_2$ and CeF_3 as a surrogate for plutonium will be shown. Salt viscosity is fit Arrheniusly against temperature to extract the activation energy of viscous flow. Similarly, the measured salts are fit according to their concentration of network-forming species to show how oligomer formation drives viscosity. Salts are tested for Newtonian behavior by measuring the viscosity at shear rates ranging from 0.1 to 100 s^{-1}.

Melting point is measured using a ThermysONE DSC & TGA with platinum crucibles. Viscosity is measured using a Modular Compact Rheometer 702e (MCR702e) with parallel plate and cup and bobber rotational methods. Density is measured using the hydrostatic method: a metal bobber is suspended by wire from a scale, lowered into the salt, and the change in mass measured. To account for uncertainty in bobber thermal expansion, three different materials are used: stainless steel, brass, and 80Ni-20Cr alloy. All measurements take place inside an inert argon glovebox.

Salts utilized in the research were provided by the Nuclear Materials and Fuel Cycle Center at Virginia Tech and from TerraPower through Los Alamos National Laboratory.

608e Development of a Lab-Scale Pulsed Column for Observing Hydrodynamic Phenomena

Philip Schonewill, Luke El Khoury, Jasmine Gacutan, Stephen Davidson, Richard Daniel, Stuart Arm

Pulsed columns are a classical technology used to perform two-phase separations – typically liquid-liquid extractions – in several different applications, including rare mineral recovery, nuclear reprocessing, and other solvent-enabled chemical separations. The general principle is to flow two immiscible liquid phases countercurrent to each other through a column and superimpose a pulsating flow onto one of the flowing liquids to encourage droplet breakup and, consequently, more efficient mass transfer between the phases. The column typically has an internal geometry to assist with the dispersion of droplets, such as a packed bed, a disk-and-doughnut configuration, or periodically spaced sieve plates. In the United States, pulsed columns have been used for nuclear reprocessing at industrial scale, but very few smaller columns are in operation that can perform solvent extractions relevant to nuclear applications.

Scaling a pulse column down contains inherent challenges arising from the impact of wall effects, the generation of small amplitude pulsations, and short liquid residence times. A lab-scale prototype system has been designed and assembled for two purposes: (1) investigation of hydrodynamic phenomena using various aqueous/organic fluid permutations; and (2) collection of performance data to assist with the design of a future lab-scale apparatus (of similar scale) for radiological service. The design and scaling of the lab-scale prototype will be discussed, along with a presentation of initial experimental data collected using simple aqueous/organic two-phase systems.

608f TRISO SNF Deconsolidation – Summary of Pnnl's 3-Year Program

Stuart Arm

Staff at the Pacific Northwest National Laboratory have progressed several projects for the Department of Energy's Nuclear Energy Office, Office of Materials and Chemical Technologies (NE-43) related to processing Tri-structural ISOtropic (TRISO) spent nuclear fuel (SNF). TRISO nuclear fuel is to be used in several commercially proposed advanced High Temperature Gas Reactors (HTGRs). In contrast to light water reactors (LWRs), HTGRs use either helium gas or liquid salt as coolant and are moderated by graphite. The quantities of graphite and the way the material is integral to the fuel means the volume of disposed SNF is significantly larger than that discharged from LWRs. Therefore, there are waste management benefits to separating the fuel from the graphite and potentially disposing of the latter as low level waste. TRISO nuclear fuel itself consists of uranium oxycarbide coated with pyrolytic carbon and silicon carbide, which complicates reprocessing the SNF. PNNL's 3-year program started with reviewing technologies applicable to deconsolidating the graphite to expose the coated fuel particles and fracturing the coatings to expose the fuel. A deep dive on the most promising technology was then conducted to evaluate its feasibility in a process plant. Finally, a flowsheet was developed incorporating the selected technology and the fuel's dissolution in nitric acid. Further research and development needs were identified throughout the program.

84b The Effects of Dissolver Geometry on Nuclear Fuel Dissolution: Fluid Dynamics and Heat Transfer.

Richard Overstreet, Andrew Ritzmann, Jarrod Allred, Neil Henson, AP Goulet

Small batch nuclear fuel dissolution experiments at PNNL's Radiochemical Processing Laboratory displayed a week to month long difference in sample dissolution rates depending on the fuel rod orientation and basket geometry. Computational Fluid Dynamics was employed to investigate the effects of internal tank features and basket geometry on acid recirculation and heat transfer. Preliminary results indicate a bias in acid recirculation flow depending on the basket features and inlet flow position. The CFD models include the effects of flow, free surface motion, and conjugate heat transfer between the various materials in the dissolver. Results of this work will clarify the significant impact of dissolver design on dissolution rates.

84d Development of a Hierarchical Multiscale Modeling Framework to Address Materials Grand Challenges for Fusion

Dwaipayan Dasgupta, Asanka Weerasinghe, Michael D. Ashburn, Omeet Patel, Sophie Blondel, Dimitrios Maroudas, Brian D. Wirth

Harnessing fusion energy is a demanding process and requires materials that can tolerate extreme environments. The key challenge for the development of fusion materials is a lack of representative laboratory-scale environments which can truly mimic the harsh fusion reactor conditions to reliably test the performance of candidate materials. As a result, fusion materials research relies heavily on computational materials modeling and simulation. Research in nuclear materials is inherently spatiotemporally multiscale in nature, and requires connecting atomic-scale phenomena, such as ion- or neutron-induced damage, to defect clustering and material damage evolution over mesoscopic scales, to leading all the way up to the engineering/reactor scale. Here, we focus on developing a hierarchical multiscale modeling framework capable of simulating the dynamical response of plasma-facing components (PFCs) under conditions representative of the environment in a fusion reactor, as an effort to tackle one of the major engineering grand challenges for the 21st century. On this front, we will discuss two representative multiscale modeling efforts related to the divertor PFC of a nuclear fusion reactor, namely, 'fuzz' formation and subsurface helium bubble evolution in the tungsten divertor.

In the tungsten divertor, experiments aiming to investigate PFC performance under continuous operation and high first-wall temperature have shown that a nanostructure with a fuzz-like morphology develops on the helium-plasma-exposed tungsten surface under conditions relevant to fusion reactor operating conditions. The formation of this extremely brittle fuzz-like structure will seriously compromise the operation and performance of a fusion reactor. To first understand and then mitigate this deleterious effect, we have developed an atomistically informed, predictive continuum-scale model that is capable of capturing the formation and the early stage of evolution of the fuzz-like complex surface morphology mediated by various underlying dynamical processes characterized by disparate spatiotemporal scales. In our modeling framework, large-scale molecular-dynamics (MD) simulation results are used to develop the datasets required to parameterize the constitutive equations for the closure of the continuum-scale model for the surface morphological response of the plasma-facing material and targeted MD simulations are used to determine the thermophysical properties of low-energy-helium damaged tungsten.

Helium is insoluble in metals and forms self-clusters and bubbles in PFC tungsten, which causes significant deterioration in the thermophysical properties of tungsten. We will present a systematic study of the subsurface bubble dynamics based on a continuum-scale, spatially distributed drift-diffusion-reaction cluster dynamics model of helium self-cluster and bubble evolution, which can predict the helium retention in low-energy helium plasma-exposed tungsten. Our cluster dynamics model, as implemented in our code, Xolotl, has been benchmarked based on large-scale MD simulation results and experimental measurements available in the literature. We will also discuss an improvement in the predictive capabilities of Xolotl incorporationg into the cluster dynamics model the interaction energy landscape between a migrating cluster with another migrating cluster or a helium bubble, leading to an enhanced helium cluster and bubble growth, respectively. To incorporate this new physics into our cluster dynamics model following the hierarchical multiscale modeling framework, a systematic parameterization of defect interaction energetics was conducted based on molecular-statics (MS) computations of helium cluster-cluster and cluster-bubble energetics.

Furthermore, we will discuss a strategy for coupling the physics of subsurface helium bubble dynamics and gas retention in PFC tungsten, as modeled by Xolotl, with the surface evolution model and computer simulator of surface morphological response to study the feedback effect of this complex interconnected physics of helium-irradiation-induced damage in divertor tungsten. Such model and code couplings will pave the way for ultimately developing a digital twin of the fusion reactor.

84e Effects of Surface Anisotropy on the Surface Morphological Response of Plasma-Facing Tungsten

Omeet Patel, Dwaipayan Dasgupta, Asanka Weerasinghe, Brian D. Wirth, Dimitrios Maroudas

Tungsten (W) is the chosen plasma-facing component (PFC) material for the divertor in the International Thermonuclear Experimental Reactor (ITER) due to its superior properties, including exceptional thermomechanical properties and low sputtering yield. Numerous experimental studies have established that PFC tungsten suffers severe surface degradation as a result of exposure to high fluxes of helium (He) and extremely high heat loads. Specifically, high fluxes of low-energy helium ions implanted in tungsten within the temperature range from 900 K to 2000 K are responsible for the formation of a "fuzz"-like surface nanostructure, which consists of fragile nanoscale-sized crystalline tendrils. This nanotendril formation is driven by stress-induced surface atomic diffusion, with stress originating from the over-pressurized He bubbles formed due to He implantation in the near-surface region of PFC tungsten, and has adverse effects on the mechanical behavior and structural response of PFC tungsten as well as on the reactor performance. Moreover, several He plasma exposure experiments have reported that during the very early stage of the fuzz formation process, different types of nanostructures are observed to appear on the PFC tungsten surface and this difference in the surface morphologies arises due to the difference in the crystallographic orientation of the plasma-exposed surfaces. Understanding how such surface patterns form has significant implications for improving the structural and morphological response of PFC materials.

Toward this end, we report here a simulation study on the effects of surface anisotropy on morphological evolution and pattern formation on a PFC tungsten surface of specified crystallographic orientation. Our analysis is based on self-consistent dynamical simulations according to an atomistically-informed, continuum-scale surface evolution model that has been developed following a hierarchical multiscale modeling strategy and can access the spatiotemporal scales of relevance to fuzz formation. The model accounts for PFC surface diffusion driven by the biaxial compressive stress originating from the over-pressurized helium bubbles in a thin nanobubble layer, which forms in the near-surface region of PFC tungsten as a result of He implantation, in conjunction with formation of self-interstitial atoms in tungsten that diffuse toward the surface. The model also accounts for the flux of surface adatoms generated as a result of surface vacancy-adatom pair formation upon He implantation, which contributes to the anisotropic growth of surface nanostructural features due to the different rates of adatom diffusion along and across step edges of islands on the tungsten surface. Moreover, the model accounts for the surface free energy anisotropy, which contributes to facet formation on the surface nanostructural features due to the difference in the surface free energy of tungsten planes with different crystallographic orientations. The surface free energy parameterization was obtained by optimally fitting the surface free energy values for different surface crystallographic orientations predicted by atomic-scale simulations based on a reliable interatomic potential; optimal adatom diffusion pathways have been computed consistently, using the same interatomic potential.

Using the model described above, we found that by incorporating surface anisotropy, we were able to predict the pyramid-type features that are observed experimentally in the surface morphology of a W(100) surface exposed to a helium plasma. We have established that the preferential diffusion of the surface W adatoms generated as a result of the surface-vacancy adatom pair formation effect, combined with the surface free energy anisotropy, are responsible for the detailed topography of the PFC tungsten surface; for the W(100) surface, these effects underlie the formation of mounds on the surface and their precise faceted features. Specifically, the adatom diffusion contributes to the formation of a pattern of distinct mounds on the surface while the surface free energy anisotropy facilitates the facet formation on the mounds, which leads to the formation of the experimentally observed pyramid-type nanostructures. In addition to the surface morphology, the surface nanostructure growth kinetics is investigated in detail and the impact of surface anisotropy on such kinetics is explained.

84f Helium Bubble Coalescence in Plasma-Facing Tungsten

Asanka Weerasinghe, Dwaipayan Dasgupta, Brian D. Wirth, Dimitrios Maroudas

Predicting surface nanostructure formation on tungsten (W) under plasma exposure conditions is of utmost importance when tungsten is used as a plasma-facing component (PFC) in nuclear fusion devices. Subsurface helium (He) transport and He bubble dynamical processes, such as bubble coalescence and bubble bursting, play crucial roles in the near-surface structural evolution and surface morphological response of the PFC material as well as in refueling/recycling of fuel. However, in spite of its importance to the dynamical response of the PFC material, a fundamental understanding of the He bubble coalescence mechanism in PFC tungsten remains incomplete. To this end, here, we report the results of a systematic study of He bubble coalescence in PFC tungsten based on atomic-scale modeling and simulations. Our PFC tungsten models consist of He nanobubbles embedded in the tungsten matrix with bubble sizes and He density relevant to the conditions of plasma exposure in nuclear fusion devices. Based on molecular-statics (MS) and molecular-dynamics (MD) simulations, we explore the governing thermodynamics of bubble coalescence over a multi-dimensional parameter space, with parameters that include bubble size, bubble separation distance, bubble pressure, bubble growth rate, and bubble location (near the PFC surface or in the bulk material), and design targeted dynamical simulations to study the underlying kinetics and characterize the governing coalescence mechanisms.

We find that the interaction energetics between two He bubbles in W can be described as an elastic interaction perturbation to a finite-width square-well potential. The width of the square-well potential has a direct correlation to the bubble pressure and corresponds to the capture radius of the two-bubble system, beyond which the bubbles interact strongly to facilitate their coalescence. In general, smaller bubbles tend to be captured by the larger bubbles due to the higher bubble pressure (He/Vacancy ratio) in the smaller bubbles. When the two bubbles are sufficiently close to each other, the defective W regions around the spherical bubble surfaces merge at the narrow W gap that separates the two He bubbles forming a defective region with a characteristic "peanut" or "dumbbell" shape. Furthermore, we find that when the W gap between the two He bubbles becomes sufficiently narrow (about one W atomic layer), the W atoms in that narrow layer are pushed away from each other to create a channel between the two bubbles, triggering the coalescence mechanism that allows He atoms to migrate/flow between the two bubbles through a stress-driven transport process. As the gap narrows, the local strain in the tungsten matrix drives the formation of Frenkel pairs to initiate the channel opening through formation of W vacancies with the corresponding self-interstitial atoms occupying sites near the channel. Moreover, we find that continuing He implantation into the bubbles increases the pressure of the post-coalescence bubble configuration, which causes emissions of ½<111> and <100> dislocation loops/segments from the bubble surface. When the two He bubbles are sufficiently large and far apart from each other (at a distance larger than the capture radius), dislocation emissions between the two bubbles cause a narrowing of the gap between them, thus accelerating the bubble coalescence mechanism.

These analyses advance our fundamental understanding of helium bubble coalescence in PFC tungsten and contribute to an atomistically derived database for further improving the predictive capabilities of coarse-grained models of near-surface structural evolution and surface morphological evolution of PFC tungsten under fusion reactor operating conditions.

84g Impact of Soret Effect on Hydrogen and Helium Retention in Plasma-Facing Tungsten Under ELM-like Conditions

Sophie Blondel, Dwaipayan Dasgupta, Dimitrios Maroudas, Brian D. Wirth, Enrique Martinez

A uniquely harsh operating environment of a fusion power plant with combined stressors makes harnessing fusion energy a challenging materials problem, and is a key factor for considering fusion to be one of the major grand challenges for engineering in the 21st century. Combined stressors like extremely high particle fluxes and heat loads that modify the plasma-facing components (PFCs) materials microstructure are a major concern. Another such critical issue, an operational challenge related to D–T fuel fusion plants, is to minimize tritium retention within materials to keep total onsite tritium inventories in the ~1 kg per plant range with environmental releases below ~1 g/year. For its many advantageous properties, including low tritium retention, tungsten is chosen as the PFC material for ITER (International Thermonuclear Experimental Reactor). One such critical PFC is the divertor, the exhaust system used for removing helium produced in the D–T fusion reaction, which is expected to undergo severe structural damages due to its exposure to helium flux under typical operating conditions. Helium is insoluble in metals and will form clusters/bubbles in PFC tungsten; these bubbles are formed close to the plasma-exposed surface and trap hydrogenic species. On the other hand, a steep temperature gradient developed in the divertor tungsten under extreme transient heat loads, characteristic of plasma instabilities that induce edge localized modes (ELMs), would promote thermal-gradient-driven diffusion of helium and hydrogenic species and, thus, influence the deuterium and tritium retention in the PFC. Taking all the aforementioned physics into account, we focus here on developing a predictive model for hydrogenic species and helium retention under fusion reactor relevant conditions.

To fundamentally understand the impact on PFC tungsten of thermal-gradient-driven diffusion, commonly called 'Soret effect', we have used nonequilibrium molecular-dynamics (NEMD) simulations to analyze the transport of He, mobile helium clusters, H, and self-interstitial atoms (SIAs) in the presence of a thermal gradient in pristine tungsten. We have found that all the species examined tend to migrate from the cooler to the hot regions of the tungsten sample, characterized by a negative heat of transport [1,2]. The findings of our thermal and species transport analysis have been implemented in our cluster-dynamics code, Xolotl, which has been used to compute temperature and species profiles over spatiotemporal scales representative of PFC tungsten under typical reactor operating conditions, including extreme heat loads at the plasma-facing surface characteristic of plasma instabilities that induce ELMs [2]. Xolotl simulations of individual species transport predict that accounting for temperature-gradient-driven species transport has a very significant effect on the steady-state species profiles in plasma-facing tungsten. We will also present the cluster-dynamics simulation results that include self-clustering of helium and/or tritium partitioning toward helium clusters and discuss the influence of the Soret effect on helium and hydrogenic species retention inside PFC tungsten, as well as plasma fueling and tritium flux at the cooling tube side. All these results and analysis are crucial for fuel cycle assessment.

[1] E. Martinez, N. Mathew, D. Perez, S. Blondel, D. Dasgupta, B. D. Wirth, and D. Maroudas, *J. Appl. Phys*. **130**, 215904 (2021).

[2] D. Dasgupta, S. Blondel, E. Martinez, D. Maroudas, and B. D. Wirth, *Nucl. Fusion* **63**, 076029 (2023).

84h Effects of Surface Crystallographic Orientation on the Surface Morphological Response of Plasma-Facing Tungsten

Omeet Patel, Dwaipayan Dasgupta, Asanka Weerasinghe, Brian D. Wirth, Dimitrios Maroudas

Tungsten (W) is the chosen material for plasma-facing components (PFCs), including the divertor, in the International Thermonuclear Experimental Reactor (ITER). While the properties of tungsten, such as its high melting point, low sputtering yield, and low tritium retention, are advantageous for its use as PFCs, a large body of experimental evidence has established that severe material degradation of PFC tungsten in a fusion reactor environment can lead to significant operational challenges. Specifically, experimental studies have shown that, under typical operating conditions expected in the ITER divertor, He implantation above a threshold of incident ion energy of approximately 35 eV causes formation of a 'fuzz'-like fragile nanostructure on the PFC tungsten surface. We have previously established that formation of a nanotendril-like structure, which would ultimately evolve to 'fuzz', is driven by stress-induced surface atomic diffusion, with stress originating from the over-pressurized He bubbles formed due to He implantation in the near-surface region of PFC tungsten. 'Fuzz' has adverse effects on the mechanical behavior and structural response of PFC tungsten as well as on the reactor performance. On the other hand, the extreme fusion environment has significantly limited the material choices for PFCs. Therefore, it is essential to mitigate this surface damage toward enabling the use of fusion reactors on the power grid where a reactor would be expected to operate continuously for many months, namely, a period much longer than the typical time scale for 'fuzz' formation, which is on the order of hours. Given this predicament, we focus here on understanding the very early stage of the fuzz formation process, during which several experiments have reported that tungsten surfaces with different crystallographic orientations exhibit different periodic patterns, including pyramidal, triangular, and stripe-shaped features. Understanding how such surface patterns form has significant implications for improving the structural and morphological response of PFC materials.

Toward this end, we report here a simulation study on the effects of surface crystallographic orientation on the surface morphological response of PFC tungsten. Our analysis is based on an atomistically-informed, hierarchically developed continuum-scale surface evolution model that can access the spatiotemporal scales of relevance to fuzz formation. The model accounts for atomic diffusion on the PFC surface driven by the biaxial compressive stress due to the over-pressurized helium bubbles in a thin layer, which forms in the near-surface region of PFC tungsten as a result of He implantation, as well as a diffusive flux toward the surface of W self-interstitial atoms. Furthermore, the flux of surface adatoms generated through surface vacancy-adatom pair formation upon He implantation is accounted for; such surface adatom fluxes contribute to the anisotropic growth of surface nanostructural features due to the different adatom diffusion rates along and across step edges of islands on the tungsten surface. Importantly, the model also accounts for the difference in the surface free energy of tungsten planes with different crystallographic orientations, which incorporate into the analysis surface free energy anisotropy effects that give rise to facet formation on the surface nanostructural features. In our study, optimal diffusion adatom pathways have been computed by atomic-scale simulation based on a reliable interatomic potential. Surface free energy parameterization was obtained by optimally fitting the surface free energy values for different surface crystallographic orientations predicted consistently by atomic-scale computations using the same interatomic potential.

Using the model described above, we attempt to obtain a fundamental understanding of the mechanisms of the stripe- and triangle-shaped feature formation that are experimentally observed in the W(110) and W(111) surfaces, respectively, when exposed to a helium plasma. Moreover, significant efforts have been made to extend the predictive capabilities of our surface evolution model to successfully predict the surface orientation dependent nanostructure formation; a detailed characterization of the predicted surface topographies will be discussed as well. We find that adatom diffusion plays a key role in determining the main qualitative features of the surface topography, such as formation of mounds or striped features on the surface, while the surface free energy anisotropy facilitates the faceting of such mounds or striped features, resulting in the full complexity of the experimentally observed surface morphologies for all surface orientations examined. In addition to the detailed surface morphology and specific morphological features for the various surface crystallographic orientations studied, the growth kinetics of the surface nanostructure is investigated in detail and the impact of surface crystallographic orientation on such kinetics is discussed and explained.

547a Materials Characterization Capabilities of the Uranium Science and Technology Center at Oak Ridge National Laboratory

Daniel Felton, Jennifer Neu, Luke Sadergaski, Avery Wood, Andrew Miskowiec, Roger Kapsimalis

The science surrounding uranium processing was mainly developed during the Cold War via weapons programs. As a result, knowledge of the fundamental chemistry and kinetics of the processing was overshadowed by a need for fast and reliable production at an industrial scale. The Uranium Science and Technology Center (USTC) will be a pilot-scale program that will study the chemical processing of uranium in both pre- and post-enrichment procedures. This will include dissolution, separations, precipitation, calcination, and halide conversion. USTC is a part of the Nonproliferation Stewardship Program and supports the goal of ensuring the US nuclear workforce maintains technical skills and capabilities. USTC will act both a R&D space for understanding fundamental science of uranium processing and as a training space for users to develop skills with sophisticated equipment and instrumentation.

547b The NSP Harmonia-Plus Lab Initiative and an Emerging Ultrasonic Separations System for Metal Hydroxide Removal

Karla Erickson

The Harmonia-Plutonium Science Laboratory (PluS Lab) initiative has a mission to grow and sustain workforce, addressing current nonproliferation needs, and posturing for emerging nonproliferation challenges with targeted fundamental plutonium S&T. Over two talks, this new initiative and examples of future projects carried out in the PluS Lab will be described. One example, illustrated in this presentation, is the development of a continuous, virtually silent benchtop ultrasonic filtration system (UltraSep) for nuclear processing applications. With increasing need throughout the nuclear complex to accelerate nuclear waste processing timelines and reduce worker dose, UltraSep is an emerging technology with the potential to be impactful in this area. The system is simple to operate, has a small footprint, and requires little to no maintenance, making it competitive with conventional membrane filtration and centrifugation technologies and potentially deployable in gloveboxes. UltraSep treats slurries to an ultrasonic standing wave, concentrating the particles and separating them from a clarified filtrate. This work focuses on utilizing UltraSep to separate metal hydroxide particles from caustic aqueous solutions. In tandem, the metal hydroxide slurries are examined using an analytical centrifuge to understand their fundamental properties and establish correlations to their behaviour in the UltraSep system. This presentation will show the results from experiments with cerium, zirconium, and depleted uranium hydroxides, highlighting both their unique particle behaviours and the efficacy of UltraSep in removing them from slurry solutions.

LA-UR-24-24850

547c Harmonia-Plutonium Science Laboratory (PluS Lab): Actinide-Molten Salt Research

Marisa Monreal

The Harmonia-Plutonium Science Laboratory (PluS Lab) initiative has a mission to grow and sustain workforce, addressing current nonproliferation needs and posturing for emerging nonproliferation challenges with targeted fundamental plutonium S&T. Over two talks, this new initiative will be described, along with examples of research areas that will be studied in the PluS Lab. This talk will focus on the actinide-molten salt research area, and will include (1) highlights of current research capabilities, with recent U and Pu results, and (2) plans for this research in the PluS Lab. In (1), novel techniques developed at LANL to study these high-temperature (400-900 oC), radioactive, and corrosive liquids will be detailed, including those used to determine macroscale properties (neutron radiography for density, drop calorimetry for heat capacity) and examine local structure (neutron scattering, Raman spectroscopy). In (2), connections to nonproliferation and ideas and initial plans for plutonium-molten salt research in the PluS Lab will be described.

LA-UR-24-24888

547d Simulating Processing of Actinides and Developing Expertise (SPADE)

Lawrence Banks, Hiroshi Saito, Saurabh Narain, Jeremy Scher, Bradley Childs

The SPADE project is an initiative within the Nonproliferation Stewardship Program addressing the growth and sustainment of technical expertise in nonproliferation applications associated with Plutonium reprocessing and recovery operations. This presentation provides an overview of SPADE with focus on coupling physics-based computational modeling and simulation with machine learning techniques to create an environment that will enable workforce development, advance science and technology, and enhance infrastructure environments to support the stewardship of nuclear fuel recycling. The progress towards a virtual simulation environment will be shared with emphasis on machine learning applications within the project.

Prepared by LLNL under Contract DE-AC52-07NA27344. LLNL-ABS-864225

547e Chemical Engineering Needs and Opportunities Relevant to Nuclear Materials Processing and Nuclear Nonproliferation

Alexander Braatz, Joseph F. Birdwell Jr., Matthew Francis

Process chemistry and engineering are major components of nuclear materials processing, from production of reactor fuels and accelerator targets, to manufacture of defense-related components, to recovery and purification of industrial and medical isotopes. Chemical engineers have played critical roles in the development, deployment, and operation of nuclear materials processes. In many aspects, chemical unit operations and technologies associated with nuclear applications are not those emphasized in undergraduate and graduate chemical engineering curricula. Traditionally, the needed skills and expertise are taught in a relatively small number of university Chemical Engineering departments, resulting in most training taking place on-the-job. The lack of available training may be justified on the basis of two, likely misperceptions: (1), all nuclear chemical processes of industrial interest are thoroughly developed and adequately optimized and (2), nuclear materials processing (particularly actinide processing) is understood in sufficient detail at all levels (fundamentals to application) that further R&D is superfluous.

As concern over the climate impacts of fossil fuels has intensified, interest in power reactor alternatives to Pressurized Water Reactors and Boiling Water Reactors has increased. New concepts require nontraditional fuels, necessitating novel chemistries and new and/or enhanced technologies. Equally as important, the growing threat of state and nonstate entities with military ambitions has increased the need for personnel trained in unit operations throughout the nuclear fuel and weapon cycles to support policy makers in nonpoliferation efforts and international nuclear safeguards. With these aims in mind, the Y-12 National Security Complex is currently standing up the Uranium Production and Weaponization Testbed (UPWT) as part of the DOE-NNSA Nonproliferation Stewardship Program. This facility is being designed to operate uranium conversion, purification, and metallization processes at pilot scales in an integrated fashion. The goal of the UPWT facility is to provide on-the-job training at an actinide processing facility while providing an S&T environment capable of addressing the design challenges associated with material processing for the civilian reactor fleet and the nonproliferation questions that the DOE-NNSA and policy makers may pose.

547f Uranium Science and Technology Center: Exploring Fundamental Science of Uranium Chemical Processing

Randy O. Ngelale

In 2020, the US Department of Energy's National Nuclear Security Administration established the Nonproliferation Stewardship Program to ensure the US government maintains the foundational technical capabilities and workforce needed to meet nuclear nonproliferation objectives. The Uranium Science & Technology Center (USTC) is one such initiative and is charged with developing and sustaining DOE's expertise in uranium chemical processing, both upstream and downstream of enrichment. Despite nearly 80 years of fuel cycle operations, the US's national proficiency in uranium chemical processing has atrophied over the past several decades, and knowledge gaps remain in specific chemical mechanisms of fuel cycle operations. USTC seeks to grow and sustain workforce knowledge thorough the advanced characterization of uranium processing by developing and operating a state of the art uranium processing and analysis center. A principal research objective of the USTC is to address science gaps that exist in front-end uranium fuel cycle processing, by:

1. Quantify process effects on the physical and chemical properties of uranium products;
2. Characterize processing effects through the parametric studies of unit operations to quantify various flowsheets;
3. Identify chemical species and chemical kinetics of unit operations

The focus of this talk will be on establishing the laboratory-scale unit operations to support workforce development and the methods currently being developed to capture ex situ and in situ chemical characterization of processing effects to better understand the fundamental chemistry of uranium processes. These empirical measurements will be validated against physics-based predictive models to quantify reaction kinetics and other process parameters.

547g The Design of Organic Simulants to Model Various Nuclear Reprocessing Fluids

Luke El Khoury, Philip Schonewill, Gabriel Hall, Gregg Lumetta

The reprocessing of irradiated nuclear fuel to extract Uranium and Plutonium from other compounds is facilitated through a liquid-liquid separation technique called the Plutonium Uranium Reduction EXtraction (PUREX) process. Since mass transfer occurs only where the two immiscible liquids meet (at the interface), a central goal when designing unit operations for PUREX is to promote a high surface area to droplet volume ratio. The three types of contacting equipment used in this unit operation are the mixer-settler, centrifugal contactor, and pulse column.

Though many designs for these devices exist, they all encourage the breakup of one liquid phase into droplets dispersed within the other to elicit a high interfacial area and, subsequently, transfer metal ions between phases. Understanding the subtle balance between flow rates, droplet size, mechanical power input, and coalescence time is highly complex and necessitates fundamental knowledge of the materials used. The most common liquids for this process are dilute nitric acid (1M - 3M) and a tributylphosphate (TBP, 30 vol%) mixture in n-dodecane. It has also been documented that key physical attributes change significantly due to increases in Uranium loading. Therefore, no system displays only a single set of physical properties because the PUREX fluids experience a wide range of changes in physiochemical properties. While it would be ideal to engage studies of the unit operations directly with the actual materials (i.e., uranium, plutonium, TBP), feasibility becomes an increasingly significant hurdle when handling radioactive materials. Therefore, economics, chemical compatibility, hazard prevention, and general intrigue orient the goal of this research to identify a selection of material substitutes for the traditional set of liquids to assist in prototyping and mock-up testing of the PUREX unit operation.

While many factors influence the performance of the unit operations and interfacial area of droplets, three vital ones are viscosity, interfacial tension (IFT), and density. Therefore, this document aims to explore and enhance the flexibility of experimental capacity by understanding, measuring, and controlling organic surrogates that span a wide array of the properties listed above. With these chemical systems, the separation equipment can be more robustly tested and scrutinized by surveying their limits while maintaining a safer and more cost-effective environment.

DISSOLUTION OF SURROGATE U-Zr FUEL USING ALNIFLEX CONDITIONS

Jarrod M. Gogolski, Robert A. Pierce, Rachael E. Fara, Thanh-Tam T. Truong

Savannah River National Laboratory, Aiken, South Carolina

1.0 Introduction

Non-aluminum clad spent nuclear fuel (NASNF) stored in L Basin at the Savannah River Site (SRS) is widely varied in fuel composition, design, packaging, and physical condition. The complexity of the NASNF inventory presents significant challenges, and technology development is necessary for successful disposition. One such fuel in the inventory is metallic uranium-zirconium (U-Zr) alloy fuel, the focus of this study. Electrolytic or nitric acid only dissolution of metallic U-Zr alloy can form insoluble zirconium oxide, which results in up to 52% loss of U to insoluble solids, and can be subject to potentially uncontrolled oxidation reactions [1, 2]. The AlNiflex process was determined to be a viable dissolution flowsheet for the U-Zr alloy fuel. Under a narrow set of solution concentrations, a combination of hydrofluoric acid (HF), nitric acid (HNO_3), aluminum nitrate ($Al(NO_3)_3$), and hexavalent chromium can safely dissolve U-Zr intermetallic alloys, keep Zr soluble, and not significantly corrode stainless steel (SS) vessels [3, 4].

A crucial aspect of the AlNiflex process that has hitherto been unstudied are the solubility conditions before, during, and after dissolution of Zr, Al, and U components. Previous research ascertained solubility conditions of Al and Zr fluorides at different anion and cation concentrations [5]. However, the dissolution of multiple NASNF fuel bundles sequentially to the same dissolver solution presents a changing dissolution environment that could cause unwanted precipitation of zirconium or uranium fluorides.

Calculations were done for five different Zr-clad U-Zr fuel bundle compositions representing a large fraction of NASNF U-Zr fuel bundles: listed in Table 1 as EBWR-1, EBWR-2, HWCTR-1, HWCTR-2, and HWCTR-3. EBWR and HWCTR denote the Experimental Boiling Water Reactor and the Heavy Water Components Test Reactor, respectively. Additionally, EBWR-1, EBWR-2, HWCTR-1, and HWCTR-2 are comprised of high U-low Zr alloy fuel and HWCTR-3 is high Zr-low U alloy fuel. Of these five, four of the calculations (EBWR-1, EBWR-2, HWCTR-2, and HWCTR-3) were evaluated experimentally for solubility behavior. The calculations simulated the sequential addition of six or seven fuel bundles to a nominal 7,000 L dissolver at SRS, accounting for acid consumption and fluoride complexation by the dissolving species. The starting conditions and process inputs (initial solution volume, HF and HNO_3 starting concentrations, and volumes of HF and HNO_3 additions) were then adjusted for lab-scale experiments and compared against data reported by Henry [5] to establish a dissolution profile. Hexavalent chromium was omitted from these tests because it was not expected to affect metal solubility and generated hazardous waste.

The goal of this study was to assess solubility conditions while dissolving six or seven surrogate fuel bundles in AlNiflex dissolvent and finish with a dissolver solution containing minimal residues or precipitates. This lab-scale study confirmed that by the subsequent additions of surrogate fuel bundles and acid, the quantity of precipitate decreases as fluoride ions form metal complexes. The exception was with the addition of the final surrogate sample for HWCTR-2. The analyzed solids were an aluminum fluoride species, indicating a slight shift in the solubility. Otherwise, the undesirable zirconium or uranium fluoride species were not observed as end-state residues.

2.0 Experimental

The starting concentrations to dissolve each simulated fuel type at production scale is listed in Table 1. These values were then directly scaled for the lab-scale tests (i.e., Bundle 1 for EBWR-1 was 1.2415 g Al and 2.3761 g Zr). In the lab-scale tests, the Al bundles containing U-Zr fuel assemblies were simulated using alloy Al-6061 and Zr-2 metal coupons. The metals were dissolved in 500 mL of prepared AlNiflex dissolvent in a Teflon reaction vessel with gentle mixing between 23 - 80 °C for varying lengths of time. It is assumed that the formation of metal-fluoride complexes, which impacts solubility, is not affected by the dissolution temperature after the solution is cooled to ambient temperature. The AlNiflex dissolvent contained HNO_3 (Fisher Chemical, Certified ACS Plus), HF (Fisher Chemical, OPTIMA grade), and aluminum nitrate nonahydrate (ANN, Alfa Asesar, ACS, 98.0 - 102.0%). The omission of hexavalent chromium was due to the expectation that the ion would not have a solubility impact for any metal fluorides, and omitting it decreased the amount of hazardous waste generated.

Table 1. Starting dissolvent conditions, representative fuel bundle additions, and calculated operating parameters for large scale dissolution. Note: The HNO_3 was added after every bundle charge and HF was added between 3rd and 4th bundle charge.

Simulated Fuel Type	Initial Dissolver Vol. (L)	Starting Dissolvent Conditions (M)			Representative Fuel Bundles (kg)			HNO_3	HF
		HF	HNO_3	ANN	Zr	Al	U	Add Vol. (L)	
EBWR-1	6300	2	2	0.5	29.8	15	81	200	100
EBWR-2	6000	2	2	0.5	35.2	15	113	250	150
HWCTR-1	5150	2	3	0.5	4.5	59	53	400	250
HWCTR-2	5450	2	3	0.5	7.4	50	52	350	200
HWCTR-3	7400	2	1.5	0.5	26.8	7.5	1.7	0	0

As expected, all dissolver solutions initially contained aluminum trifluoride (AlF_3) precipitate. The starting solution included both the precipitate and supernate. The reaction vessel was periodically heated using heat tape and checked with a Teflon-coated Type K thermocouple. Note that some of the metal samples were dissolved at room temperature. After each surrogate fuel bundle completely dissolved, a 14 mL sample was withdrawn and stored in a centrifuged tube for later addition of uranium oxide (UO_3). After the sample was removed, water, concentrated HNO_3 and/or concentrated HF were added to the main solution along with the next surrogate fuel bundle based on the dissolution calculations. As with the Zr and Al metal, the amount of liquid was scaled for the lab-scale (i.e., between 3rd and 4th bundle charge for EBWR-1, 4.58 g HF and 7.92 g of HNO_3 was added). The process was repeated until all the surrogate bundles were dissolved. The set of 14 mL samples were observed for at least a fortnight to allow solutions to stabilize. Since the initial solutions did not account for uranium oxide (UO_3) dissolution, 13.8 M (?) nitric acid was added directly to each centrifuge tube prior to the addition of UO_3. The tubes were vigorously mixed by hand and then gently heated in a water bath set to 50 °C for 6.5 h (with caps loosened to prevent water loss via evaporation).

3.0 Calculations for Simulated Bundle Dissolutions

Although there is a wide range of U, Zr, and Al concentrations contained in the NASNF inventory, the dominant quantities of bundles could be approximated by the four bundle compositions listed in Table 1. These provide the basis for an initial set of solubility calculations as a function of bundles dissolved. To clarify, the calculations were done for large scale batch dissolution of fuel bundles in the H-Canyon dissolver and then adjusted to approximately 500 mL lab-scale experiments using surrogate fuel bundles (metal coupons). Constraints from the solution solubility, process chemistry, and of the operating facility were considered during the dissolution of the surrogate fuel bundles. For example, F^-:Zr ratio needed to be greater than or equal to 5 to prevent an uncontrolled exothermic dissolution of U-Zr. For most types of fuel bundles, this leads to needing one addition of HF during operations and limiting the number of bundles dissolved (5 - 7 bundles of fuel per nominal 7,000 L dissolver batch) such that the system is free of solids at the end of the dissolution batch [5]. Although the AlNiflex process has been shown to dissolve up to 0.5 M Zr without the formation of precipitate, these calculations limited the final Zr concentration to less than 0.35 M [3]. It is acceptable to precipitate AlF_3 during the dissolution of the first several bundles. This is due to an increase of solubility of AlF_3 as Zr and Al metal, which complex F^-, dissolve. A low F^- concentration needs to be avoided to prevent zirconium fluoride (ZrF_4) precipitation because ZrF_4 is highly resistant to dissolution [3].

The AlNiflex initial dissolution conditions used 2 - 3 M HF to facilitate Zr and Al metal dissolution, 1 - 3 M HNO_3 for acceptable Zr, Al, U-Zr, and U dissolution rates without excessive SS corrosion, and 0.5 - 1.0 M ANN for initial corrosion control without excessive fluoride complexation [3, 6]. The final volume in the dissolver was maximized (up to 7,400 L) to have as much of the bundle submerged at the start to enable shorter dissolution times with the initial volume adjusted to accommodate additions of HNO_3 and HF between bundle charges. When necessary, the calculations assume the addition of 15.8 M HNO_3 between every bundle charge and the addition of 28.5 M HF between the third and fourth bundle additions. Although the documented AlNiflex flowsheet initially uses 1.0 M ANN for corrosion control, the calculations suggest that 0.5 M ANN would reduce the number of HF additions required, an operation considered difficult for the operating facility. Table 1 provides the calculation parameters for the representative bundle types, which yield the data compared in Figure 1 and the target concentrations used for the simulated bundle dissolution - solubility experiments.

It was assumed that 6, 3.5, and 4 moles of H^+ were consumed per mole of Zr, Al, and U, respectively, and that 4, 0.9, and 0 moles of F^- complexed per mole of Zr, Al (between 0.5 to 1.3), and U, respectively [5, 7-10]. The calculations were compared with the Al and Zr solubility data reported by Henry, also shown in Figure 1. The values from the literature in Figure 1 were used to predict whether an Al or Zr precipitate can be expected. If the final two simulated bundles were outside of the Figure 1 constraints (noted by the enclosed area) and indicated the presence of a precipitate, the process conditions (initial or in-process) were modified. The solubility calculations provided the basis for subsequent dissolution and solubility experiments. Equation 1, based on data from the literature, defines the end state of the dissolution process (S). Precipitation is avoided at greater than 0.8 M H^+ and when S is between 0.5 and 1.6 [3, 11]. Note S is unitless and denotes when precipitants form.

$$S = \frac{F^-_{total} - 4 \cdot Zr}{Al} \qquad (1)$$

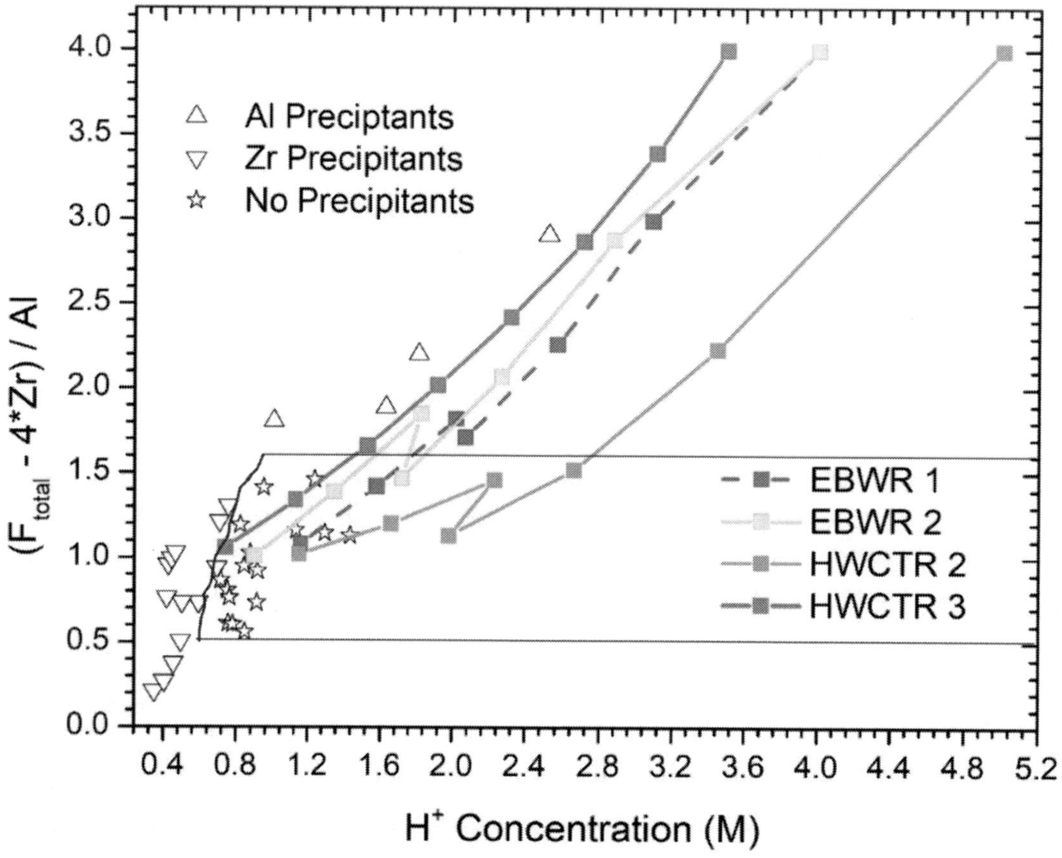

Figure 1. Effect of acid concentration on solution stability at 25 °C. The hollow points are data adapted from Henry [5]. The solid points were obtained from estimated solubility calculations and represent the solubility effect from the 'bundle' additions. The lines are added for clarity to denote the surrogate fuel bundle additions, starting from initial solution solubility (S = 4).

The calculations of the moles of H^+ consumed and F^- complexed account for the molecular weights of each metal (Al = 26.98, Zr = 91.22, U = 238). As a result, Al is the biggest consumer of both H^+ and F^-, despite the Zr masses in EBWR-1, EBWR-2, and HWCTR-3 and the masses of

U in EBWR-1 and EBWR-2. Interestingly, the calculations for the dissolution of HWCTR-3 fuel bundle indicate the process can be successfully completed for 7 bundles of SNF without any supplemental additions of HNO_3 nor HF. This is possible because of the low H^+ consumption and by using the full volume of the dissolver for the starting conditions.

Conversely, HWCTR-2 requires a lower starting volume to accommodate the need to add large volumes of acid for Zr/U metal dissolution across six bundles while maintaining the H^+ concentration low throughout processing for corrosion control. As a result, the total quantity of HNO_3 and HF in the dissolver at the start is lowest for HWCTR-2. The changes in the solubility parameter (S) from bundle to bundle are lower for HWCTR-2 because the changes in the dissolver concentrations are masked by the large HNO_3 additions between bundle charges. Conversely, HWCTR-3 shows larger changes in the solubility parameters between bundles because there are no HNO_3 or HF additions to replenish the dissolver solution.

4.0 Results and Discussion

Figure 2 shows pictures of the centrifuge tubes containing aliquots taken after the dissolution of metal charges from the surrogate EBWR-1, EBWR-2, HWCTR-2, and HWCTR-3 solutions prior to and after addition of uranium oxide. Figure 1 indicated solids would still be present after the first several surrogate fuel bundle additions of the lab-scale amounts. However, the precipitate dissolves into solution (partially or fully) as the Al and Zr coupons dissolve. Therefore, the solids mostly disappeared by the 4th charge of metal coupons. Close inspection of the centrifuge tubes revealed minuscule quantities of residue in every tube except Tube 7 of HWCTR-3. Unfortunately, the quantity of residue was too small to analyze, but it could have been a small quantity of ZrO_2 from the surface of the Zr metal coupons prior to dissolution. Separate experiments indicated that a ZrO_2 coupon (stabilized using magnesium oxide) will dissolve within the Alniflex solution but extremely slow.

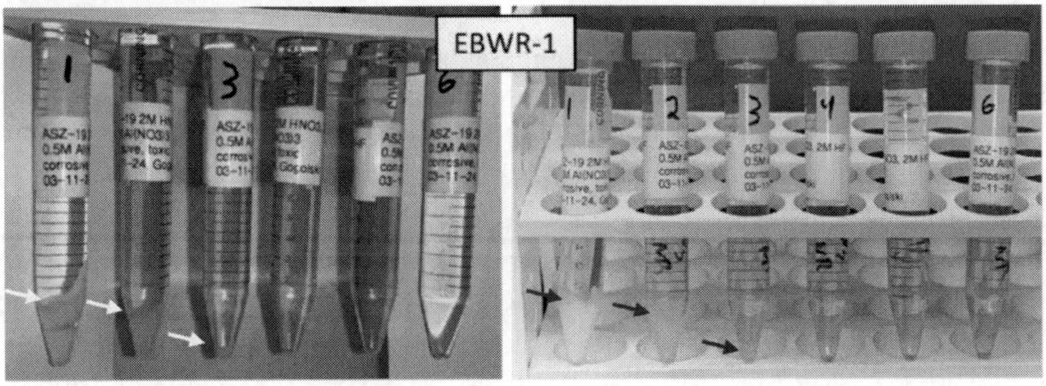

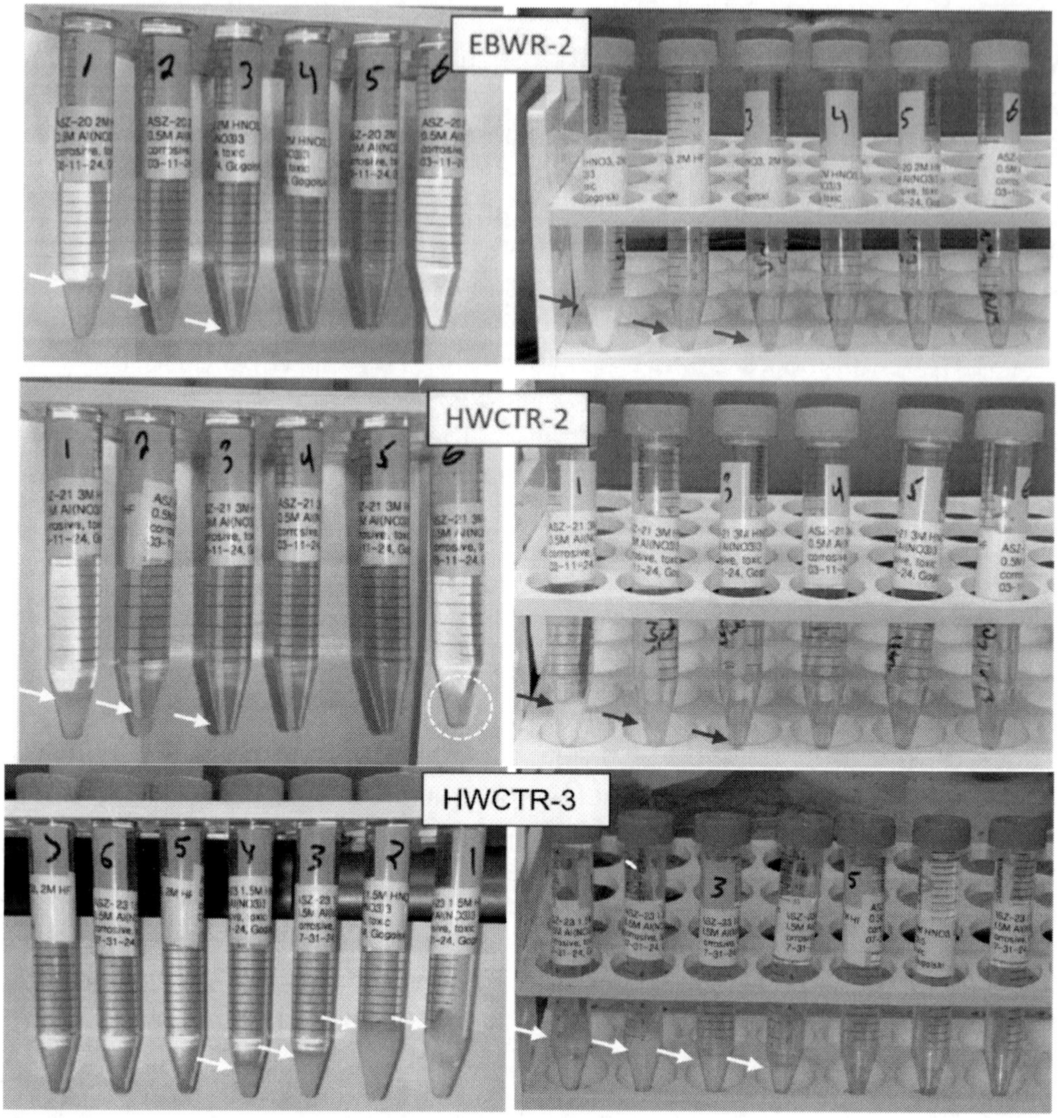

Figure 2. Surrogate fuel bundle dissolution solids for EBWR-1, EBWR-2, HWCTR-2, and HWCTR-3 before uranium addition (left) and after uranium addition (right). Arrows are a visual aid to denote level of solids in each tube.

The volume of solids visually decreased with each simulated bundle addition. HWCTR-3 might be an exception as Tube 2 appears to have more solids than Tube 1, and then the quantity of solids after Tube 2 gradually decreases. EBWR-1 had precipitates in Tubes 1, 2, and 3 (beyond residues). EBWR-2 had precipitates in Tubes 1, 2, and maybe 3 (beyond residues). HWCTR-2 had solids in Tubes 1, 2, and 6 (beyond residues). HWCTR-3 has solids in Tubes 1 - 4, trace solids in Tubes 5 and 6, and no solids in Tube 7. Although the precipitation behavior exhibits trends that are consistent with the predictions of Figure 1, the differences indicate that the assumptions in the calculations for acid consumption and fluoride complexation could benefit from measurements specifically associated with metal dissolution experiments using the AlNiflex dissolvent. The presence of solids in HWCTR-2 after the 6th bundle addition was not expected.

A picture of the solids at the bottom of the digestion vessel at the conclusion of the experiment is shown in Figure 3. An analysis of the solids by x-ray diffraction (XRD) identified as hydrated aluminum fluoride hydroxide or AlF_3.

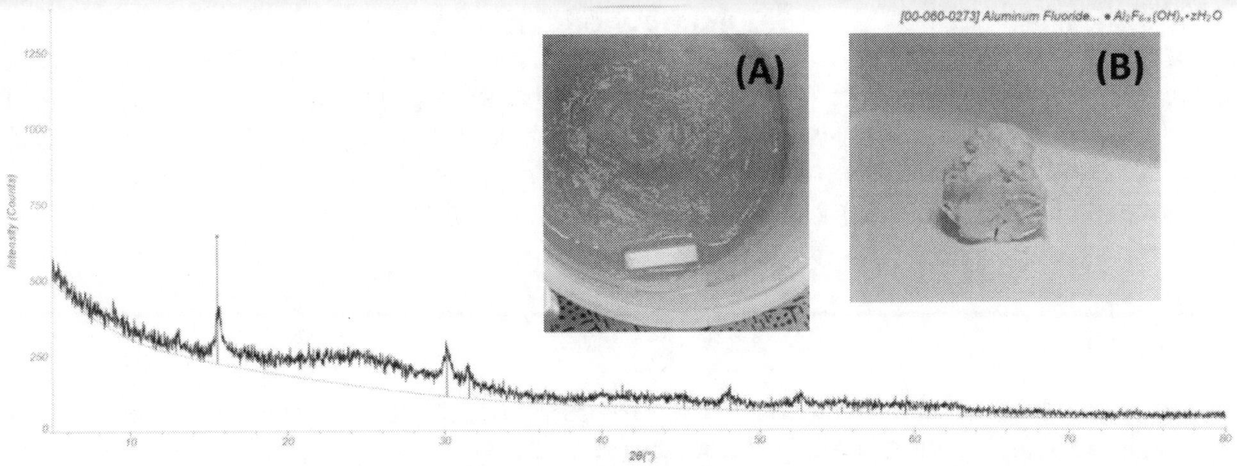

Figure 3. XRD pattern of solids after 6th simulant bundle addition for HWCTR-2. Inset shows solids after container emptied of dissolvent (A) and after a month of being in a desiccator (B).

It is worth noting that previous studies of the AlNiflex process focused on high Zr-low U fuels like HWCTR-3 [3, 11]. Since these fuels are high in Zr but low in Al and U, they do not consume large quantities of HNO_3. The low U-high-Zr fuels are high in Al and U, which consume large volumes of acid and complex significant concentrations of F^-. The later addition of UO_3 and HNO_3 did not visibly change the quantity of solids in each tube, as is evident in the right-side images in Figure 2. It is important to note that these solubility studies targeted 0.3 - 0.35 M Zr due to the volume constraints of the H Canyon dissolver. This is a lower concentration than the initial AlNiflex flowsheet studies which dissolved 0.5 M Zr. However, at 0.5 M Zr, the final solutions were stable for only a few days before precipitation of solids occurred [5, 11]. The solutions produced in these studies have been stable for weeks without the formation of additional solids.

5.0 Conclusions

Calculations were done using a set of acid consumption, fluoride complexation, and solubility assumptions based on the literature and framed according to constraints associated with dissolving Zr-clad U-Zr fuel stored in Al bundles in H Canyon. The calculations simulated the additions of six or seven surrogate fuel bundles with HNO_3 and HF as a proof of concept for U-Zr fuel dissolution in H Canyon. Precipitation behavior was noted during the dissolution of Zr and Al metal coupons under AlNiflex conditions during the dissolution of six to seven simulated fuel bundles, and then the effect of dissolved U was simulated through the addition of UO_3 solids and HNO_3. The observed solubility behavior was consistent with the calculations and the literature. Most importantly, after dissolution, the solutions with 0.3 to 0.35 M Zr were stable for several weeks without the formation of precipitates. These results continue to confirm the feasibility and viability of the AlNiflex flowsheet for the disposition NASNF fuel bundles at H Canyon.

6.0 Acknowledgements

This work was produced by Battelle Savannah River Alliance, LLC under Contract No. 89303321CEM000080 with the U.S. Department of Energy. Publisher acknowledges the U.S. Government license to provide public access under the DOE Public Access Plan (http://energy.gov/downloads/doe-public-access-plan).

7.0 References

1. R. P. Larsen, et al., "A study of the explosive properties of uranium-zirconium alloys" (ANL-5135). Argonne National Laboratory. (1954).
2. E. S. Occhipinti, "Processing of power reactor fuels twelfth quarterly progress report" (DP-546). Savannah River Laboratory. (1961).
3. V. P. Caracciolo, and F. G. Rust. "AlNIFLEX process for dissolving zirconium-uranium alloy in 304L stainless steel vessels," *I&EC Process Design and Development*, **5**(*4*), 364-368 (1966);
4. R. A. Pierce, "Chemical dissolution flowsheets for processing uranium-zirconium alloy fuels in H-Canyon" (SRNL-TR-2023-00311). Savannah River National Laboratory. (2023).
5. H. E. Henry, "Recovery of uranium from zirconium-uranium alloy" (DP-826). Savannah River Laboratory. (1963).
6. E. S. Occhipinti, "Reprocessing of power reactor fuels, third quarterly progress report" (DP-319). Savannah River Laboratory. (1958).
7. N. K. Chaudhuri, and R. M. Sawant, "Stability constants of the fluoride complexes of actinides in aqueous solution and their correlation with fundamental properties" (BARC/1997/E/022). Bhabha Atomic Research Centre. (1997).
8. E. M. Vander Wall, E. M. Whitener, and D. L. Bauer, "Dissolution of Zirconium and Zircaloy-2 in Nitric-Hydrofluoric Acid Mixtures at Reflux Temperatures" (IDO-14497). Phillips Petroleum Co. (1959).
9. R. A. Pierce, "Uranium Metal Dissolution in the Presence of Fluoride and Boron" (WSRC-TR-2003-00500). Savannah River National Laboratory. (2003).
10. R. Klein. "Dissolution of zirconium in nitric-hydrofluoric acid mixtures," *Corrosion Engineering Section*, **53**(*4*), (1996);
11. H. D. Harmon, "Penetrate-leach dissolution of zirconium-clad uranium and uranium dioxide fuels" (DP-1424). Savannah River Laboratory. (1976).

486b Extraction Model Development and Enhancements for the Amuse Code

Laura Maggos, Candido Pereira

Recent interest in developing simplified solvent extraction processes for recycle of high-assay low-enriched uranium (HALEU) has motivated the expansion of solvent extraction modeling capabilities in the Argonne Model for Universal Solvent Extraction (AMUSE). The monoamide extractant N,N-di-2-ethylhexyl-isobutyramide (DEHiBA) is under study for the selective separation of uranium from the transuranic elements and the other irradiated nuclear fuel components. Distribution ratio (D value) models for key nuclear fuel components (including U, Np, Pu, Am, Tc, and Zr) are being developed for solvent extraction with DEHiBA in AMUSE to support development of uranium extraction flowsheets. The updated AMUSE code has been used to design spent fuel treatment flowsheets that use DEHiBA as the extractant. Additional capabilities added to AMUSE to improve performance and applicability to different processes will also be described.

The submitted manuscript has been created by the UChicago Argonne, LLC as Operator of Argonne National Laboratory ("Argonne") under Contract No. DE-AC02-06CH11357 with the U.S. Department of Energy. The U.S. Government retains for itself, and others acting on its behalf, a paid-up, nonexclusive, irrevocable worldwide license in said article to reproduce, prepare derivative works, distribute copies to the public, and perform publicly and display publicly, by or on behalf of the Government.

486c Opportunities for Radiochemical Separations of Radioisotopes from Molten Salt Reactors

Jisue Braatz, Kristian Myhre, Hunter Andrews, Joanna McFarlane

Molten salt reactor (MSR) technologies are attracting considerable interest for commercialization due to their inherent safety features and efficient energy production. Besides generating reliable and clean energy, MSRs hold the potential to harvest various high-value radioisotopes. However, the feasibility of directly recovering these isotopes from the fuel salt remains uncertain. This study investigates the behavior of different radioisotopes within a MSR environment and explores the theoretical and practical aspects of electrolytic processes in high-temperature fused salts. Electrolytic processes, extensively used for molten salt processing, present a promising approach for the direct extraction of isotopes from the fuel salt. We review the general principles of fused-salt electrolytic processes, including electrowinning, electrorefining, electroplating, and electroextraction, and discuss several experimental considerations. Additionally, we summarize published experimental results on isotope extraction via electrolytic methods. Our findings highlight critical factors for the effective recovery of radioisotopes from MSR-relevant molten salts through electrodeposition. This research provides insights that are crucial for advancing the commercial viability of MSR technologies and optimizing the recovery of valuable radioisotopes.

486d Impact to Savannah River Site Processing Facilities from Combining Fuel Spent Nuclear Fuels with High-Level Waste

Dan Lambert, Dylan Rodene, Wiersma Bruce, Fabienne Johnson, Corey Martin

The Defense Waste Processing Facility (DWPF) at the Savannah River Site (SRS) near Aiken, South Carolina, processes and immobilizes the legacy waste from the production of nuclear weapons materials (along with smaller quantities of nuclear isotopes, Pu for space missions and research materials). The Salt Waste Processing Facility (SWPF) will process approximately 100 million gallons of soluble radioactive waste and produces two additional waste streams that are processed through DWPF. In addition to SRS produced nuclear waste, Spent Nuclear Fuel (SNF) from reactors outside SRS is being stored in a former reactor disassembly basin at SRS. Changes to waste composition require a thorough facilities evaluation as unexpected changes to processing conditions can result in costly equipment failure and schedule impacts.

In this paper, we discuss the studies that were performed to assess the impact of adding the neutralized SNF with SRS high-level waste (HLW). The SNF comes from various US reactors along with a number of foreign reactors. SRS used aluminum cladding for its fuel and targets whereas the SNF has a variety of claddings including aluminum, stainless steel, Zircaloy, Hastelloy, etc. These difficult to dissolve SNFs all require more aggressive dissolution flowsheets, requiring electrolytic dissolution and HF. The neutralized slurries can be very different than historical SRS HLW, many high in fissile uranium and plutonium.

This paper discusses the impact of combining the non-SRS SNFs with SRS sludge and the impacts on processing in DWPF and SWPF. The emphasis in the paper will be the impact of added fluorine and fissile actinides (particularly U-235, Pu-239 and Pu-241) on downstream processing facilities. In order to combine the non-SRS SNF without negatively impacting sludge processing, the SNF will be dissolved in nitric acid without actinide recovery. The solution will be neutralized with sodium hydroxide and added to high level waste (HLW) tanks for processing in DWPF and SWPF.

486e Overview of Radiochemical Processing at Oak Ridge National Laboratory for Plutonium-238 Production

Kristian Myhre, Joseph Hough, Joseph McVeigh, Adam Parkison

Deep space exploration missions require reliable power. Plutonium-238 is the radioisotope of choice for the US National Aeronautics and Space Administration (NASA) to provide power to deep space probes, such as rovers, in the form of radioisotope thermoelectric generators. Oak Ridge National Laboratory (ORNL) is establishing constant rate production of plutonium-238 heat source material to support future NASA deep space exploration missions. The production life cycle of plutonium-238 involves a number of radiochemical and nuclear material processing steps. These steps include purification of neptunium material, chemically transforming dissolved neptunium nitrate into oxide powder, pressing mixed neptunium oxide with aluminum powders into cermet pellets, fabricating metal-clad reactor target bodies containing the neptunium cermet pellets, irradiating the neptunium targets in a reactor with a high flux of neutrons, and radiochemically processing irradiated targets in a hot cell facility to result in a plutonium-238 oxide product stream and neptunium nitrate stream for recycling into new targets. The processing of waste produced throughout the entire production life cycle into disposal forms is also crucial. This presentation will provide an overview of radiochemical processing at ORNL for plutonium-238 production and discuss new research directions that are being explored to improve efficiency.

418a Nuclear Nonproliferation Workforce: Keeping Pace in a Changing World

Jennifer Ladd-Lively

The Nuclear Nonproliferation community has a need for a sustained technical workforce across numerous domain expertise to continue to support its mission. In order to sustain this workforce, we must of new people in the pipeline as well as senior experts that can transfer knowledge prior to retirement. With the current growth in the nuclear industry and the rapidly emerging technologies, it is difficult to get people into the nuclear field and to keep them there. I will discuss Oak Ridge National Laboratory's efforts within this topic.

418b *Grand Challenges in Nuclear Nonproliferation Related to Irradiated Fuel Processing*

Gregg Lumetta, Brienne Seiner, Harris Eldridge, Kevin Lyon

The expertise relied on to execute the United State's nuclear nonproliferation mission was developed primarily through experience in operating industrial scale facilities within the U.S. nuclear weapons complex. However, such facilities used to process irradiated fuel for recovering plutonium ceased operations at the end of the Cold War. Hence, over the last 35 years this first-hand knowledge of nuclear materials processing has atrophied, with most workers from that era no longer available to address nuclear nonproliferation challenges. This issue has become a major concern for the National Nuclear Security Administration (NNSA), leading to the establishment of the Nonproliferation Stewardship Program, and in particular Project Athena, which is specifically directed at processing of irradiated fuel. Athena is a multi-laboratory effort that seeks to reinvigorate and grow U.S. expertise in the recovery of nuclear materials from irradiated fuels and targets, with a focus on the nuclear nonproliferation mission. Given that the U.S. no longer processes irradiated fuel, three science and technology grand challenges have been established by Project Athena which revolve around process scalability, process vulnerability, and process maturation. This presentation will discuss the nature of these grand challenges.

418c Processing Challenges Impacting Completion of Radioactive Waste Processing at the Savannah River Site

Dan Lambert

The Savannah River Site now has all the processing facilities needed for processing the high and low activity radioactive waste needed for emptying the 43 remaining active waste tanks and closing down the processing facilities when this mission is complete. But the site also wants to process as much spent nuclear fuel to combine with the SRS waste before shutting down these facilities. So SRS has the challenge of finding processing conditions suitable for simultaneously optimizing H Canyon fuel dissolution, sludge and salt preparation, the Defense Waste Processing Facility, the Salt Waste Processing Facility, the Saltstone Processing Facility and the recycle/effluent treatment facilities at processing rates that haven't been achieved to date in spite of the limited lag storage. Any slowdown or shutdown of one of these facilities leads to the idling of all of these facilities. The challenges of optimizing the processing of these facilities and the likely bottlenecks will be discussed.

418d Nuclear Fuel Cycle, Chemical Engineering Opportunities and Challenges

Monica C. Regalbuto

The United States is pursuing an aggressive plan to combat climate change by reducing carbon emissions in which nuclear power is set to play a vital role. Accelerated development of advanced reactor designs introduces evolving nuclear fuel cycles that deviate from the commercially established light water reactor. As new fuel cycles and technologies emerge, both challenges and opportunities are presented. This presentation will cover the importance of addressing fuel cycles front-end and back-end emerging concepts to ensure the successful deployment of advanced reactors and the unique role chemical engineers play.

277a Invited Talk: Dr. James Willit, DOE NE

James L. Willit

This invited talk with panelists is aimed at discussing the US nuclear fuel cycle status and opportunities for the nuclear chemical engineering profession. This is a "workshop"-like format event aimed at discussions with speaker, panelists, NED members, and audience.

214a Raman Spectroscopy for Monitoring Particle Size in Complex Slurries for Nuclear Waste Management

Rupanjali Gurprasad

This work investigates the potential of Raman spectroscopy as a tool for determining solid particle properties within slurries, with an emphasis on particle size. Slurries, common in industries such as chemical, pharmaceutical, and nuclear waste management, present challenges in monitoring and understanding solid-state properties during processing. Raman spectroscopy, traditionally utilized for qualitative and quantitative material analysis, shows promise in providing insights into solid properties within slurries. The research proposes a novel approach of using Raman spectroscopy to assess particle size by monitoring the attenuation of a solution phase signal in the presence of added solids. Simulated high-level waste slurries resembling those at Hanford are employed, with silica solids of varying particle sizes introduced into an alkaline solution. Continuous monitoring of Raman signal attenuation allows for the modeling of particle size prediction. While complexities may arise in multicomponent systems, this approach holds particular promise for crystallization systems, where understanding particle properties like size is critical for quality control. Employing Raman spectroscopy as a singular tool for discerning particle size, alongside its conventional role in compound quantification, offers a cost-effective strategy for investigating critical particle properties within slurries.

214b A Full-Scale Mixing and Transport Feasibility Study Using Non-Radioactive Solid Simulants in a Mock-up Electrolytic Dissolver to Enable the Processing of Spent Nuclear Fuels

Dylan D. Rodene, John M. Pareizs, Jacob Mellon

As part of a directive to disposition all of the spent nuclear fuel (SNF) remaining in storage at the Savannah River Site (SRS). The H-Canyon processing facility at the SRS is being adapted to readily process legacy SNFs. H-Canyon will dissolve and neutralize the SNF within the facility and discard the acid-soluble portion of the SNFs to the Concentration, Storage and Transfer Facilities (CSTF), where the discards will be blended and mixed with other tank waste from across the SRS. The blended waste will form sludge batches to be processed in the Defense Waste Processing Facility (DWPF) and turned into a vitrified-glass waste-form.

A risk assessment was performed for the disposition and handling of the legacy SNFs, which includes a group of legacy non-aluminum SNFs (NASNF). Several risks associated with electrolytically dissolving zirconium (Zr) clad NASNF were identified. Of which, a significant risk for the buildup of solid residue in the electrolytic dissolver over time was emphasized, as ~85% of the Zr processed is likely to be converted into insoluble residue that is mainly composed of ZrO_2. The risk for this task is elevated further owing to the fact that the legacy electrolytic dissolver (last run in the 1980's) was recently found to contain a significant amount of legacy insoluble solids residue residing in the bottom of the dissolver. This indicates that the legacy residue removal campaigns were not successful, and the residue was never removed from the dissolver bottom post-processing. Therefore, SRNL was tasked with developing a full-scale mock-up rig to evaluate the feasibility of the handling and removal of the dissolver residue by using non-radioactive simulants and to provide operational recommendation for the removal campaigns.

This work will touch on the development and use of non-radioactive simulants for the feasibility of performing insoluble solids removal campaigns post-dissolution in an electrolytic dissolver. Emphasis will be on the density and morphology dependence of the simulant, as well as mixing performance and removal efficiencies of the simulants at various solids loading. The evaluation suggests that the removal campaigns will result in a high likelihood of success in managing the buildup of solids within the dissolver. Some opportunities were identified to optimize solids transfer and water usage by configuring the operating parameters of the mock-up dissolver tank. The cycling of the spray ring (the method was used to mix the tank using water as the motive fluid) in the dissolver bottom, water level in the tank, and the outlet suction height were some of the key parameters identified to influence the mass transfer and mixing of the simulants in the tank.

214c *Enhancing Regulatory Compliance: Assessing Hazardous Waste Standards with Low-Code Tools*

Heather B. Sabella, Courtney Bottenus, Brett Simpson

The Hanford Site, a former nuclear production complex, generated significant waste over its 30-year operational period, resulting in 56 million gallons currently stored in 177 tanks. This waste, comprising over 1,800 chemicals, poses substantial challenges for management in terms of storage, treatment, and disposal. Oversight is managed by the Department of Energy (DOE), Office of River Protection (ORP), with support from Washington River Protection Solutions (WRPS). Regulatory authority is delegated by the Nuclear Regulatory Commission (NRC) and Environmental Protection Agency (EPA) to Washington State Ecology and Washington State Department of Health during decommissioning. Understanding the waste's characteristics, such as ignitability and reactivity, is crucial for compliance with Washington State regulations under the ORP mission.

As waste is sampled and studied, classifications may evolve, necessitating periodic reviews. The most recent review involves analyzing over 2 million rows of sampling data and more than 50 report files (average page count: 133, average file size: 4.4 MB). To automate parts of this review, Data Scientists and Chemical Engineers at Pacific Northwest National Laboratory (PNNL) are exploring Tableau and Microsoft Power Platform products. Tableau facilitates data cleaning and visualization of sampling data to compare with regulatory standards. Meanwhile, the Power Platform extracts metadata (e.g., author, publish date) and content details (e.g., keywords) from reports to establish a structured document architecture. This architecture addresses gaps in sampling knowledge and offers historical context for waste code designations. This paper explores how incorporating these low-code tools can enhance efficiency and usability, providing comprehensive insights to support informed decision-making in waste management strategies.

214d Evaluating Metallic Nd and Pr Interactions with Molten Chloride Salts Using Spectroscopy and in-Situ Electrochemical Spectroscopy.

Stephanie Castro Baldivieso

After the nuclear fuel (i.e., UO_2) is used inside the reactor, the fuel matrix is characterized by a variety of fission products (FP). Many of the FP (i.e., rare earth elements, alkali/alkaline earths, actinides) in the used nuclear fuel can be recovered via nuclear reprocessing technologies. In pyroprocessing, the used nuclear fuel is electrochemically dissolved into a molten chloride (LiCl-KCl) eutectic salt mixture in the electrorefiner. Due to the different applied potentials, uranium is initially reduced onto an inert cathode. However, many of the remaining FPs accumulate in the melt and are challenging to recover by an inert electrode, especially the rare earth elements (Nd, Gd, Pr, Sm) due to their multivalent oxidation states, disproportionation reactions ($Nd^{3+}+Nd^{0}=Nd^{2+}$) and undergo dissolution in the electrolyte. Due to these challenges, FP recovery is inefficient and leads to the constant discarding of the molten chloride salt which in turn generates more waste. Additionally, the presence of rare earth elements and other fission products in the molten salt electrolyte changes the physical and chemical properties of the electrolyte, which impacts the efficiency of the recovery of uranium, and the lifespan of the molten chloride salt. To improve the recovery efficiency of the FP, specifically rare earth elements, the fundamental interactions between rare earth elements in the molten chloride salt and their metallic form: LiCl-KCl-$NdCl_3$, LiCl-KCl-$PrCl_3$ with pure Nd and Pr metals were investigated at 773–873 K. The kinetics pathways of the disproportionation/dissolution reactions occurring at the interface between the rare earth metal and the molten chloride salt were investigated via UV-Vis and Raman spectroscopy. The electrochemical transitions of Pr^{3+}/Pr(s), $Nd^{2+,3+}$/Nd(s) were evaluated using an inert W electrode at 773–873 K using cyclic voltammetry, chronoamperometry techniques coupled with in-situ UV-Vis spectroscopy to characterize their kinetic pathways and contrast the results obtained by chemically reacting the metallic rare earth and the molten chloride salts.

214e Dead-End Filtration of Dilute an-107 Hanford Tank Waste Supernatant at Reduced Temperature

Christian Alvarez, Jarrod Allred, Edgar C. Buck, Richard Daniel, Jarrod E. Turner, Amy Westesen, Reid Peterson, Matt Landon

The U.S. Department of Energy's Hanford Site is home to 56 million gallons of highly radioactive legacy waste stored in underground tanks. Washington River Protection Solutions (WRPS) manages the tank farms and has deployed the Tank-Side Cesium Removal (TSCR) system designed to treat the tank waste supernatant by removing suspended solids and cesium via filtration and ion exchange unit operations, respectively, before the supernatant is immobilized by vitrification for long-term storage. The Direct Feed Low-Activity Waste test platform developed and operated by Pacific Northwest National Laboratory consists of bench-scale purpose-built systems that mimic TSCR unit operations and is used to demonstrate the filtration, ion exchange, and vitrification of batches of aqueous tank waste supernatant provided by WRPS. The dilution and filtration steps are crucial in maintaining ion exchange cesium removal performance and robustness. The most recent demonstration began in 2023 with the dilution of 9 L of AN-107 supernatant from >9 M Na to 5.5 M Na with raw process water sourced from the Columbia River followed by reduced-temperature filtration utilizing a Mott Media Grade 5 sintered metal dead-end filter at 16 °C with a nominal flux rate of 0.065 gpm/ft^2 and a transmembrane pressure not to exceed 2 PSI – parameters and materials prototypic of the full-scale TSCR dilution and filtration processes. Filter maintenance procedures employed by TSCR including back-pulse to reverse fouling when the transmembrane pressure reaches 2 PSI, filter cleaning with 0.1 M NaOH, and clean water flux measurements pre- and post-filtration to compare filter performance before and after the test were also executed on the bench-scale system. The unique physical and chemical properties of the AN-107 supernatant resulted in filter fouling and eventual plugging that was resistant to back-pulsing and cleaning methods that had been successful with previous tank waste filtration campaigns, requiring the use of oxalic acid to restore filter performance. Multiple fouling mechanism functions were fit to filtration data to identify the predominant blocking regime in the filtration of AN-107 supernatant. Total alpha analysis conducted on permeate samples showed transuranic (TRU) content exceeded the 0.1 μCi/g threshold established by the Department of Energy to define TRU waste by a factor of 10. Scanning electron microscopy and scanning transmission electron microscopy examination showed the presence of fluorophosphates and Mn-Fe phases in the dewatered solids samples collected from back-pulses and scraped from the filter. The filtration test suggests the Mott Media Grade 5 filter may be prone to quick and sudden plugging in the TSCR processing of AN-107 supernatant that has not been given sufficient time to settle and may require cleaning and maintenance protocols that diverge from methods that have been effective with previous tank wastes.

AUTHOR INDEX